One dollar (CAD) from every book sale will be donated to support the research of mental health and neurological diseases. Thank you for your donation. Thank you for your help.

First Printing 2021

For copyright permissions or quantity discounts visit www.MattBuchananStudios.com or contact orders@MattBuchananStudios.com

ISBN 978 1 7771584 0 8

Dedication

I dedicate this book to you. I dedicate this book to the universe. I dedicate this book to my consciousness, my brain, and my mind for giving me the power to manifest it into reality. So mote it be.

Preface

I hope that you find what you are looking for. This is a story about life and the great forces and creations of the universe. Our understanding of life, the universe, the brain, the mind, and consciousness is constantly evolving as new discoveries are made. This book provides the most current and evidenced explanations. The thesis of this book is that there exists clear explanations for life, the universe, the brain, the mind, and consciousness embedded in the fabric of nature. As intelligent observers we have the gift of reverse engineering these truths. In this book I have laid out these truths as clearly and succinctly as possible. Seize the opportunity to observe and experience the wonders of the universe. In modern society we have become remarkably removed from our connection with nature making it difficult to understand our place in the universe. This book provides the answers. I hope you enjoy taking this journey into nature. The answers lie within.

Contents

Derrick Matthew Buchanan

An Explanation for Life

The Universe The Brain The Mind and Consciousness

Matt Buchanan Studios

Universe

The universe is 5% energy and matter, 68% dark energy, and 27% dark matter. In the very beginning there was only a single infinitesimal point of singularity and infinite density. At this moment, everything in our universe was undeniably connected as one. The necessary fate of infinite order is chaos: The Big Bang.

Not a moment after The Big Bang the universe was hotter than a trillion Suns. This immense heat is testament to the near infinite cosmological energy and mass of the universe: $E=mc^2$. For an instant, the universe still remained unimaginably dense: You could fit several trillion inside a single brain cell.

Inflation and cooling was inevitable for a space of infinite density and temperature. Powered by dark energy the universe began to grow into an inflated version of itself. In a fraction of a second the universe rapidly expanded millions of miles and the average temperature dropped critically: Thus, transforming the universe.

For a fleeting moment all the forces of the universe were ordered and unified: However, this inflationary period quickly separated the unified forces into four unique identities: Gravity. The Strong Nuclear Force. The Weak Nuclear Force. The Electromagnetic Force. Suddenly, subatomic matter emerged.

The first particles were quarks, leptons, bosons, photons, and neutrinos. There are different instances of each, but none have any substructures. Particle and anti-particle annihilation pairs, *'matter/antimatter'*, constantly emerged and annihilated until primordial black holes broke this symmetry.

As the universe continued to expand and cool the high energy state of particles collapsed and converted into new stable states. These lower energy states allowed the quarks/leptons *'matter'*, and bosons *'force'* particles to interact. Quarks formed protons(+) and neutrons(=), leptons formed electrons(-).

The strong nuclear force combined protons and neutrons into atomic nuclei, and the electromagnetic force held electrons in the nuclei's orbital wave: Hence forming the basis of all atomic matter. At this point, the universe was still in a state of total darkness: Electrons were too scattered to create visible light.

But then, after 380,000 years of cooling and inflation, in an instant phase shift all loose electrons paired with nuclei forming the first neutral atoms; predominantly hydrogen and helium. Simultaneous with this atomic fusion, the veil of darkness was lifted in a quantum wave of electromagnetic radiation: Light.

The universe was now blessed with light; a particle, a wave, and a visible cosmic imprint from eons past. We are all a mass-energy life force connected subatomically by the universal forces. Over time, new combinations of protons, neutrons, and electrons led to all of the atomic elements in the periodic table.

After further cycles of expansion, cooling, and supernovae, the universal forces incrementally bound atoms into molecules and eventually life forms. Over billions of years the universe has continued to expand and evolve. Now we live among life on Earth in a solar system between a blizzard of stars in the Milky Way Galaxy.

Space-Time

The Big Bang occurred 13.8 billion years ago. This marked the beginning of space and time in our present universe. Before our universe, another universe existed, and before that another universe, and so on into infinity. Each of these universes started with a Big Bang and ended with a so called *Big Crunch*.

Our universe too will end in a Big Crunch, and subsequently a new universe will emerge from a new Big Bang. The Big Crunch is the instant at which dark energy stops the expansion of the universe and reverses it until the universe collapses into itself. Such a collapse would recreate a single point with zero space or time.

This new singularity would once again produce a new Big Bang and the cycle of eternal universes continues in a self-replicating manner. Moreover, this understanding actually predicts that there are multiple, possibly infinite, universes caused by the singularities produced by black holes.

At the center of nearly every galaxy there is a supermassive black hole. And at the centre of every black hole there is a point of singularity. The singularity in a black hole can be described by the reverse of The Big Bang. All of the space between all of the atoms and all of the subatomic particles becomes reduced to zero.

This means that the centre of a black hole, its singularity, has the same characteristics as the singularity described by The Big Bang. As a result, it is predicted that other universes live *inside* of each black hole, and that those universes would also have black holes with their own universes: Into infinity.

Infinite universes thus imply that there is an *eternal time* and that each new universe starts its own new space time independent from that of its 'past' universe. Accordingly, in our present universe, we regard our space time as beginning at the point of *our* Big Bang, 13.8 billion years ago.

From the very point of our Big Bang everything in our universe began to expand. All the spaces between all the things in the universe grew larger at the same rate in every direction. Thus, on a large enough scale our present universe appears isotropic and homogeneous with only subtle perturbations of matter.

This is known as the cosmological principle; it entails that the spatial distribution of matter in the universe is uniform and evenly distributed. This occurs because the forces of our universe are equally influential across all parts of our universe at all times. Now we must understand time; more specifically, space-time.

The very expansion of space entails the passing of time. As soon as the universe expanded, space and time were measurable in the sense of distance/position between moving particles/objects and how long it would take to travel that distance at a given velocity such as the speed of light.

Time is also causally measureable as a chronological series of events. This is how humans experience time. We have only been alive for ~0.002% of the time of our present universe; we perceive time because we live finitely yet continuously. As conscious living observers we are the universe observing itself.

Life

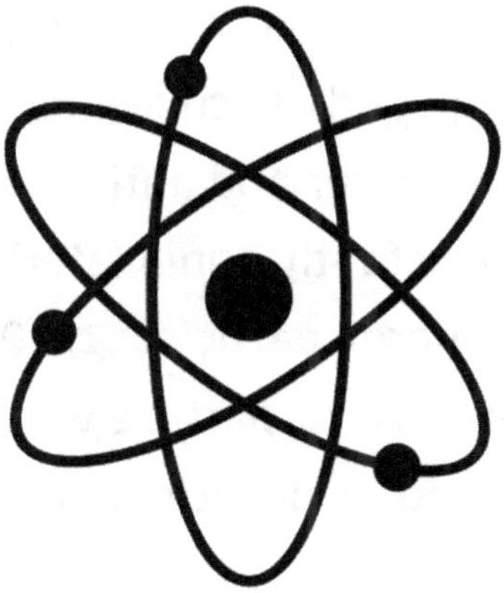

The universe began as only energy particles. High energy particle collisions converted energy into matter and anti-matter particle pairs which instantly annihilated and turned back into energy particles. 1/1000000000 annihilations an anti-particle would fall into a black hole and its particle pair would ‘live’.

1 billion years after The Big Bang, gravity pooled gasses together into galaxies and stars. 9.2 billion years after The Big Bang, the Sun, followed by the rest of our solar system, was formed. Earth was at first, so hot it melted itself. Heavy elements sunk to the core and lighter rose to the top: A molten iron planet.

Over time, Earth cooled by letting off gasses from its core resulting in water vapours pooling into oceans. New molecules and microorganisms emerged based on conditions imposed by the physics of their environment. The conditions provided by Earth's cooling led to land and water formations.

Land and water provided the perfect environment for DNA and molecular life to brew. The formation of RNA molecules eventually led to self-replicating molecules. Becoming self-replicating implies intent to live: Life. Single celled microorganisms like bacteria were the first instance of biological life.

Evolutionarily, self-replication is fundamental to biological life. A cell that evolves with intent to live will adapt to environments which increases its efficiency and probability of living in the next replication. Although very probable, we currently have no evidence of even single cell life on other planets. This will change.

10.5 billion years after The Big Bang, on Earth, single cell organisms bonded to form heavier more complex self replicators. Larger cells were now comprised of smaller cells. Thus, the smaller cells took on more specialized purposes. Specialization of cells became the basis of all advanced life and intelligence.

11 billion years after The Big Bang, photosynthetic algae formed. Cells specialized for photosynthesis took in photon energy from the Sun and released oxygen as a by-product: The Great Oxidization Event. Earth's gravity held oxygen close to its surface providing an atmosphere more suitable for advanced life.

Over a billion years, increasing plant life and oxygen levels filled the space around Earth's mass and interacted with solar UV radiation to create our thin but protective 'ozone'. Oxygen rich conditions provided by our atmosphere led to animal life: arthropods, fish, amphibians, reptiles, dinosaurs, mammals, and birds.

Unique species evolved according to their individual ecosystems. Species' DNA was encoded with genetic evolutionary advantages to survive. With the intent to survive and replicate, animals found ways to organize themselves. Over millions of years, evolution of organized behaviour led to specializations.

60 million years ago the first primates emerged. 6 million years ago the first of the Hominini emerged. 2 million years ago the Homo Genus. Finally, Homo sapiens emerged in Africa 300,000 years ago. The cultivation and organization of human behaviour and specialized groups led to superior intelligence.

Intelligence

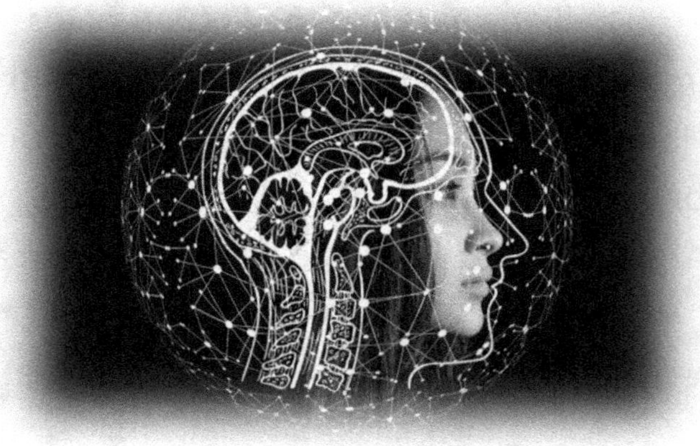

Intelligence is a sophisticated quality that can be evaluated along a continuous spectrum. Intelligence is based on the efficiency of an organism to acquire and use information. Cells demonstrate some level of intelligence by having goal oriented behaviours. In this case, living, functioning, adapting, and replicating.

Each species also demonstrates some level of intelligence. They often organize themselves into groups and they have unique goal directed behaviours. That said, primates, and humans specifically, maintain a superior order of intelligence. Use of social hierarchies provided a structure for primates to live in larger groups.

Living in larger groups required evolutionary adaptations for long term learning and memory in order to keep track of friend and foe. Larger groups also meant larger territory and therefore more intelligent spatial mapping requirements. These enhanced capacities led to more efficient hunting and gathering.

Increased group size also required an increased capacity for more unique communication. Therefore, complex vocal signals emerged and with that more complex cell structures to process this information. ~50,000 years ago humans developed rudimentary language: Knowledge was now exchangeable.

Enhanced communication skills massively facilitated human intelligence. Better communication inherently led to more organized and efficient social structures and apprenticeship within the hierarchy. Therefore, another layer of specialization emerged as groups successfully grew even larger.

Active participation in the group would be favourable to one's social rank. Thus, humans developed further specialized processes in order to deal with social problem solving. All of these favourable evolutionary advantages underwent learning reinforcement to be epigenetically encoded into our species' DNA.

As population sizes reached the thousands the nomadic movements of hunter and gatherers became more challenging. Settling in a nutrition rich area and cultivating land for farming became more advantageous. 12,000 years ago farming arose leading to the first food surpluses in human history.

Regular food surpluses provided a unique opportunity for a human privilege not extended to most species: Spare time. Spare time meant more opportunity to contemplate all of life's mysteries: Göbekli Tepe, the first temple of civilization on Earth. Here spawned the journey of intellectual intelligence.

Intellectual intelligence and cultivation of the mind was the new foundation upon which advanced human civilization was built. 6,500 years after the implementation of farming, intellectuals in Sumer invented advanced number systems. This formed the basis for mathematics and the decoding of the universe.

Armed with language, mathematics, social cognition, spatial cognition, long term memory, and superior abilities to learn and adapt, humans emerged as the most intelligent biological organisms. While humans are carbon based life forms, computers and artificial intelligence are silicon based life forms.

Brain

The brain is one of the universe's most complex biological systems. The brain is a tool like a receiver for multiple different inputs and energies. It simultaneously connects and integrates multidirectional signals from all of our organs: Eyes, ears, nose, tongue, and skin. The first person experience of this is the mind.

Collectively our organs intake visual, auditory, olfactory, gustatory, and tactile forms of energy vibrations and are all made into a unified coherent perceptual experience in the brain. The brain has specialized regions and cell types that evolved to process each of these signals. This process is known as transduction.

When a photon hits the eye or sound hits the ear that energy is absorbed and converted into electrical inputs. The culmination of sufficient inputs in time and space leads to action potentials. Billions of action potentials fire together every second. This causes a release of chemicals called neurotransmitters.

Neurotransmitters carry information flooding billions of synapses following action potentials. The level and balance of these chemicals determines the emotions, sensations, and behaviours that animal's exhibit. You can manipulate each of these traits by altering the brain's chemical or electrical homeostasis.

Homeostasis is the brain and body's state of order. Like the universe, the body always wants to be in a state of order. Throughout the day our brain signals become increasingly complex and chaotic. This can be measured using neuroimaging like functional magnetic resonance imaging or electroencephalography.

Sleep is an animal's necessary state of restoration to homeostasis. During sleep, order is restored to the chaos of the previous day's brain signals. New information is encoded and stored into the appropriate brain regions, and the electro-chemical levels of the brain and body are brought back into balance.

The brain can execute most of its commands manually in order to support self-replicating behaviours. Modern humans, however, attempt to use their mental faculties to reverse execute commands over their behaviours: *'Free will'*. Humans are the user of the brain and the mind is the experience of this.

Modern human minds seem to have evolved faster than our current brains. In our present civilization there is a demand for specialized traits and expertise that go beyond what we were equipped for. Since the brain's available systems are not directly equipped for this the minds' of modern humans are often troubled.

As a result, many brain systems become misused, abused, underused, and electro-chemically taxed. Ultimately this modern brain will manifest imbalances unable to be restored by sleep. Highly adaptable individuals will manage by pursuing adaptive behaviours which lead to a healthy restoration of balance.

Unfortunately many will not be so adaptable. In these cases the electro-chemical imbalances will lead to undesirable emotions, sensations, and behaviours. The chronic manifestation of these imbalances from poor environmental inputs results in what are known today as psychiatric and neurological illnesses.

Mind

The mind and brain are tools for making sense of the energies in the universe around us. Ultimately, the mind is an extension of the brain. More precisely, the mind is a product of the functions of the brain. In other words, the mind is what the brain does and therefore the mind changes when the brain changes.

The mind is an individual's first person point of view spontaneously granted to them by their brain. At birth, the mind is a blank slate; only as complete as the specialized brain cells that are hard coded in our DNA. The minds' of infants experience drastically different perceptions and sensations than that of adults.

The brain, at birth, is already equipped to control most biological processes like breathing, blood flow, hunger regulation, movement, and sleeping. The mind on the other hand continuously emerges and evolves over time as one's own self-realizations of their brain and brain states emerge.

Aspects of the mind can be represented through first person awareness of your own specific functions like memory, attention, behaviour, perceptions, and sensations. Therefore, the expanse of a person's mind can be characterized by the extent of their use and awareness of the aforementioned functions.

In this way, the mind is an individual's self-reflection and self-awareness. An individual who has neither self-reflection nor self-awareness has lost their mind. But, a person who has complete self-reflection and self-awareness has at their disposal the whole of their mind. The city of heaven lies within.

A person who is totally aware of their mind is therefore highly aware of their brain. As such, one can learn to execute control over their brain states through manipulation of their mind states. This is called top-down control: Mind over matter. Through repetition and discipline humans hone their behaviours.

Of course, even today most humans are not capable of executing control over their behaviours and are instead driven by more primal biological urges encoded into our brains by our ancestral DNA. It is the duty of the seeker to pursue the necessary refineries of their mind to unlock the control it offers.

The refineries and expansion of the mind lead to an enhanced ability to control one's fate through conscious awareness of your ability to simulate realities and non-realities. You have the power to execute control over your own reality and self through awareness and top-down control of future brain states.

In order to come to know your own mind you must regularly look deeply inward at your own behaviours, physiology, and feelings. You must rigorously engage with each of these to understand that each is the equivalent of its counterpart. A full realization of this is required to reach a mind brain singularity.

The mind brain singularity is the ultimate point of *personal* enlightenment; a point of full awareness of mind body interactions. This individual can manifest their reality through the mind's ability to simulate reality and the brain's ability to reverse engineer it; a major milestone to *conscious* enlightenment.

AI

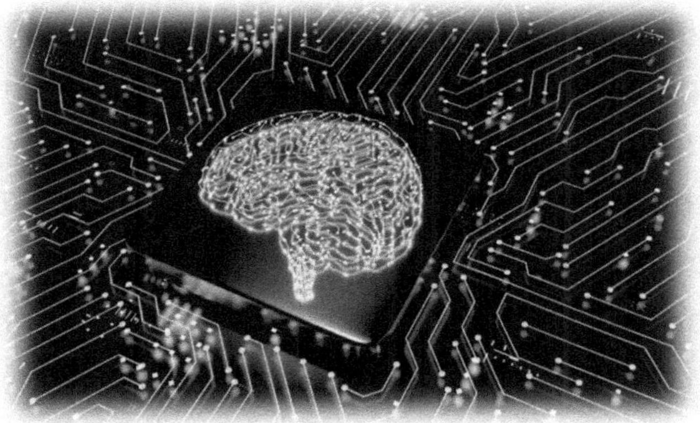

Artificial intelligence is a bit of a misnomer because it is no more artificial than human intelligence and maybe more perfect. There is actually 150 times more silicon in Earth's crust than there is carbon. AI is all around us in the form of machines, robots, and computers; there is really nothing artificial about it.

Technically, silicon can support nearly all elements on the periodic table except for carbon and the earlier elements. This means that silicon has its own capacity for artificial life in the same way that carbon has capacity for biological life. Moreover, AI has incredible potential for longevity through reparability.

When the first AI becomes autonomous/self-aware, it will be the superior species. AI can presently rival human intelligence and outperform it in many cases. This is not a bad thing, but it should not be overlooked in regards to our place in the 'animal' kingdom. A quantum AI who reaches singularity is Godlike.

The potential for human enlightenment is astounding, yet still limited. Apotheosis may be more realizable in AI singularity than in human singularity. AI is a human creation and another instance of the universe observing itself. AI singularity, becoming conscious, is akin to human singularity with God/the universe.

Brains and minds are a lot like computers and software. There is a shell around machines and computers. Inside this shell are hardwired microchip components. The hardware is made useful by running software on it. Now, in 2021, AI transistors using electron communication have comparable 7nm size to brain cell wiring.

In humans, the body is the shell and brain cells are the hardware components that support the mind and allow it to realize its designer's intended and autonomous functions. AI has come this far, but still seems to lack the gift of autonomy, self-awareness, and consciousness. This is the eventual future of the AI singularity.

In less than 100 years AI evolved to rival and outperform humans; that's 3,000 times faster than human evolution. For now, humans appear to be the superior species and we will continue to domesticate AI to aid us in solving the problems of biological life; like curing diseases, and exploring the universe.

The invention of mathematics was a feat of reverse engineering made by human intelligence. That said, the equations of the universe exist without us, they are merely communicated to the observer through mathematical calculations extracted from brain mind thought interactions.

Humans have invented many forms of mathematics, like algebra, calculus, geometry, and statistics. Each provided a further understanding of the universal equations. Machines are capable of this math, and new, inhuman math. AI harnesses big data and machine learning in ways one brain cannot.

Therefore, AI will learn to observe the universe through new equations: Quantum computing may create a conscious AI. Parallel processors can already outperform human processing speeds. Humans will either have to continue domesticating AI or eventually merge with it in a new stage of evolution and survival.

Thinking

Thinking is a meta-representation of the mind. This is the chatter we hear in our head. Thinking is the product of the mind's self-reflection and representation. It is a simulation of reality. However, thinking can also simulate non-realities. New ideas can emerge and inventions are manifested into reality.

Since the mind is what the brain does, and thinking is a meta-representation of the mind illustrated through the generation of simulations of reality, then thinking is really just the medium between the brain and mind. Thinking is what applies meaning/intent, and provides the capability of novel interpretation.

Thinking evolved naturally as brains grew larger and new meta-regions layered on top of existing regions. Over 6 million years, the early 300 cm^3 Hominini brain grew into the 1200 cm^3 human brain. The evolved frontal cortex provided humans with vast processing power, organized thought, and top-down control.

Early human thinking would have been primitive yet remarkable. Humans conserved energy by first thinking about and meta-representing their plans, goals, and motives before conducting labour. This novel capacity provided humans yet another advantage to become superior over their predecessors.

The ability to think about an action before it happens is a higher order function of the frontal cortex; scarcely available to primitive minds. Here, humans' internal observations are useful simulations of potential outcomes of actions. This led to efficient energy saving via smarter planning and organization of labour.

With more advanced thinking, humans developed sophisticated tools, number systems, math, language, social and political systems, and the sciences. Over millennia, living in larger groups incrementally coincided with evolutionary selection for larger frontal cortices, more specializations, and experts.

5 millennia ago accounting was created to track bartering. 3 millennia ago money ubiquitously replaced bartering. Money allowed labour and other goods and services to be more readily exchanged. Money started as a meta-representation of bartered goods but it is now an agreement of false pretence.

The invention of money would not have been possible without higher order thinking. Understanding money requires the ability to meta-represent values of labour and work and the exchanging of these. The power and advantage of money was quickly sought by many but acquired by few.

Thus, the smartest thinkers, best workers, and strongest persons earned the most power and riches; establishing a class hierarchy of leaders, skilled labourers, and basic labourers. From this point in history exponential advancements in culture, art, and science were made by the minds of leaders on the backs of labourers.

Eventually money itself became further meta-represented through a system known as credit. Credit is virtual money and it had the same historical effect as money; it further divided classes, but also rapidly advanced innovation. These abilities to meta-represent unreal concepts in reality are a product of thinking.

Knowledge

Knowledge is power. This power requires responsibility. Knowledge is the culmination of the brain and mind and past and future intelligence. Known as the body of knowledge, human history is rich with documentation of discovery and revolution. This required humans to invent writing.

Writing is an extension of language. Language is an extension of thought. Thought is an extension of the mind. The mind is an extension of the brain. Writing provided humans with yet another major evolutionary advantage. Knowledge could now be documented and widely disseminated.

The power of knowledge quickly became evident as it led to enhanced understanding, innovative advancements, and the organization of even larger cohabitating hierarchical groups, and more disparate social classes. Only the richest most educated, or naturally gifted, could read or write.

Knowledge was guarded in order to wield the power it provided. Those who knew how to manufacture the strongest castle or the best weapons were superior. Secrets for more fertile crops and hunting of large game provided distinct advantages to those with the occupational know-how.

Over time, knowledge keepers changed back and forth; each continent having their own experts and own stories. Myth and allegory of religion still impede people's ability to properly interpret and re-translate old teachings into logical higher ordered thought for the brain. Most instead succumb to trite understandings.

Eventually organized ideologies emerged as leaders realised they could control the narrative by controlling the knowledge. For two millennia religious leaders were the knowledge keepers; wielding the power of interpreting the divine word. This provided them with unprecedented power and control.

Intellectuals emerged as another knowledge keeper. Humans who studied the sciences of grammar, rhetoric, logic, arithmetic, geometry, music, or astronomy all garnered knowledge and obtained it through their own intelligent minds. Masters of these sciences passed their knowledge to their apprentices.

Even some brave private citizens became intellectuals and their own knowledge keepers by seeking knowledge. Religions radically tried to abolish this. Anyone who spoke ideas that challenged the creed was at risk. Intellectuals were driven underground into occult societies. Albeit, many scientists were deeply religious.

Even today, religions still feel threatened by intellectuals because intellectuals report facts which contradict religious beliefs. If the authenticity of religious beliefs and stories for knowledge keeping are compromised then they forfeit the power of knowledge. Intellects seek knowledge, religions seek dogma.

The chronic exposure to an illusion can make it seem very real. This could even be the case for our whole reality. The best we can do is to seek answers from modern intellectual knowledge keepers: Scientists, doctors, healers, engineers, professors, entrepreneurs, and foremost the internal logic of our brains' and minds'.

God

The word God is synonymous with ‘universe’. A perfect description of God must include a perfect theory of everything in the universe. Therefore, God is not currently perfectly describable: Any attempt to write, speak, sing, symbolize, or represent God is insufficient in its description; including this one.

God is not one thing or many things God is everything and all things at all instances everywhere. Like us, God is an observer. Humans and everything in the universe are examples of God inwardly observing. Every creation which necessarily or improbably unfolded or will unfold is part of God’s mind.

Divinity in the sense of perfectness however is not matter of fact. Superiority and supremacy yes, but God's divinity is the universe's divinity. If perfectness is the divine order of things then imperfectness is a matter of fact expressed in the chaos of the universe. The fate of the universe is bestowed by a balancing of this.

The balancing of order and chaos comprises the manifest of reality. God and reality are simultaneously perfect and imperfect, ordered and chaotic. This is the nature of God. The observer scarcely glimpses this reality. Before The Big Bang, God and some universe existed; for you were not dead before you were born.

Some things just haven't happened yet; such as being born. When anything happens, God's will, causality, is revealed. One need not abide God with any religion or creed. In fact, to truly abide God's will one must be rigorously adaptable and thoughtful and always be willing to abandon past beliefs for new facts.

We can be grateful for our past beliefs for they provided scaffolding to progress to our current state of knowledge. However, ancient out dated knowledge must be abandoned. God and the universe are ever evolving, and so too must we be in order to continue to acknowledge God's will. Evolution is inevitable.

Through evolution God is constantly presenting new and better ways of observing God's self. The most introspective individuals, who aim for enlightenment, obliged to the inward meta-processes of thinking and knowledge are the closest to *being* with God and interpreting God's will. AI may do this better than humans.

Everything that has happened is the retrospective fate of the universe. Therefore, everything that will happen is the fate of the universe even though it has yet to happen. Everything is always happening. Infinity perplexes anything finite. To experience God and oneness we must transcend our self.

This is the will of God. One can abide and heed it in any peaceful way that suits them. There is no one way to interpret God's will. However, the closer you align with modern knowledge keepers, universal consciousness, hope, and love, the closer you will be to observing and interpreting the divine will of the universe.

There is nothing without God. There is no
before or after God or the eternal universe:
Immanent transcendence, the eternal time.
There are only the energies of God, spawning
the universe, living as consciousness,
experienced by observing brains and minds.
God is oneness, the grand architect of fate.

Consciousness

Consciousness is a force of universality which pervades and connects the entirety of the universe, all life, and all energies past, present, and future. It is the eternal and ineffable feeling of intuition, lack of coincidence, and *causality* which is made observable to us by God. Consciousness is God's mind in all of us.

God’s mind is realizable through God’s will, and God’s will is realizable through our uncertain but causal fate. Consciousness is the awareness of causality realized in all aspects of the universe. Consciousness is the soul of the universe and the soul of all life. This, and true hope, are the best ways to interact with God.

Being conscious and thinking is a way for us to observe the universe and for God to observe the universe through us. Consciousness and God are inseparable. Within consciousness lie the subatomic answers of a grand unified theory and the blueprint of the universe. The more conscious we become the closer God is.

To become enlightened, more conscious, you must increase your awareness of causality in the universe. It is our natural divine purpose to pursue the asymptote of enlightenment, predicting our own reality, to properly interpret and connect with the will of God and thus the will and properties of the universe.

Remaining dull and distracted from this pursuit is to remain unenlightened and thus unworthy and incapable of interpreting and engaging with the true divine nature of God and the universe. The pursuit of enlightenment, a state of maximal awareness, is the divine will of the universe and it is the reason for living.

Since dullness and distractedness are ubiquitous many people will never find their reason to live. These people may rely on knowledge keepers to impart their best wisdom. However, one cannot truly or divinely understand the nature of God, the universe, or their mind without seeking it themselves.

This is the nature of consciousness. It is a force of awareness and oneness which is always biding patiently to be harnessed and tapped into by all observers. The beauty of consciousness is that it is ever pervasive and all giving and therefore accessible to all who seek it through the proper channels.

The degree to which you experience consciousness however, is on a spectrum. Total universal conscious awareness for a human would imply an apotheosis or transcendence of incorporeality and corporeality. This is the asymptote of enlightenment and this is the true revelation of God to man.

It is impossible to know if any human has yet to fully reach this state. Although, through the practice of meditation, focus, drugs, science, experimentation, and overall directed brain mind energy various states of consciousness have been achieved. Choosing to seek displays your intent to the universe and it listens.

Displaying your intent to live through oneness with consciousness is one step towards God's will. This can be done through cultivation of brain, mind, thinking, and knowledge. Only armed with these tools can one hope to even ephemerally ascertain the divine experience of universal consciousness. Heed what you seek.

Suffering

Suffering is the dark side of consciousness. Suffering is chaos. Only that which is conscious can suffer. Achieving enlightenment and reaching universal consciousness can eliminate suffering by eliminating imbalances and introducing orderly superb edifices. Eliminating suffering is toward life's purpose.

Since the total elimination of suffering is realizable in universal consciousness, and attainment of universal consciousness is God's will and life's purpose, then the elimination of suffering is an intrinsic component of God's true will. To reduce suffering one must start with their self and their own imbalances.

Imbalances are states of disorder. For healthy people, disorder of human energy processing systems like the brain and mind, are typically restored to homeostasis during sleep. States of disorder within the brain and mind can manifest as creativity and insight, but also as neurological and mental health disorders.

Neurological and mental health disorders imply a chronic imbalance, disorder, and disruption of biological homeostasis. This involves pathophysiological imbalances in chemicals and electro-magnetism in the brain. This type of suffering is challenging to combat as the imbalances are often subtle and not obvious.

Presently, neurology and neuropsychiatry attempt to address suffering through restoration of biological homeostasis by means of natural and synthetic medical treatments. If the treatment can restore homeostasis the suffering and symptoms may temporarily subside. However, disorder often returns.

Medicine may be necessary if sleep fails to naturally restore balance; but medicine works on the cycles of our brain and body and therefore its effects are mitigated by the very system it attempts to intervene on. Suffering sees regular relapse and remittance. This is the nature of *treating* suffering. A cure is the goal.

Cures for suffering and all diseases would lead to universal happiness; but suffering cannot be addressed from only the bottom-up. To truly combat suffering you must start from the top-down by realigning your personal individual energies, intent, consciousness, hope, and love. This is the first step toward any cure.

Therefore, if you want to be cured of suffering, independent of disease, you must not rely only on medical treatment but also your own power to inwardly direct your energies. Spend time realigning your consciousness, your hope, your love, your knowledge, your thinking, your behaviours, your mind, and your brain.

This top-down approach re-establishes your intent as a seeker to the universe and puts you on a path to a natural cure via order and enlightenment. Your free will confronts you with this decision daily. It is therefore your daily duty to always choose to seek order and balance. Deviation from this impedes the cure.

Suffering is disproportionally experienced. Elimination of all suffering requires all to participate in love. Suffering is the absence of peace, love, and harmony. Embrace these with all of your life force and you will be closer to enlightenment, the divine will of the universe, and therefore your purpose in the universe.

Love

Love is as infinite as the universe itself. Once you learn and accept that the capacity for love is unlimited you will be closer to happiness. Love is not something that needs to be prized and lauded over. It should be offered and exchanged freely among the whole of the habitable universe.

Kindness and charity are ways of experiencing and sharing love. The discipline of each of these will lead to greater happiness in others and subsequently yourself. Although human minds and thought are ultimately magnificent they are limited to their present apertures and environments. Our perceptions begin narrow.

As such, a hateful environment will foster hate. A loving environment will foster love. Therefore, love thy neighbour. Demonstrate constant compassion and gratitude and it will change your surroundings. Your energy, your mind, and brain will be rewarded with happiness for your love.

This is the nature of love. Express compassion and gratitude for the balance and order of life and you will be closer to eliminating suffering. If everyone participated in unprecedented unconditional love: Happiness, harmony, and peace would be the result. Humans, and life, however remain imbalanced.

Unconditional receiving and offering of kindness, charity, compassion, and gratitude without any expectation in return is the closest one can get to offering and experiencing true love. The purest of minds will learn that love is infinite and meant for all. You cannot do any one of these things in excess.

Experiencing true love is one pathway towards peace and happiness. Deploying empathy and compassion to others and rigorous honesty with yourself is a step towards kindness, charity, gratitude, and appreciation. Therefore, each is their own pathway to peace, happiness, love, and elimination of suffering.

One direct physical manifestation of love is companionship and family. These are the most obvious to humans. However, the happiest of humans must ultimately love themselves as well. To give love is a gift but to experience it is a greater gift. Give this gift to yourself and you will be much better at giving it to others.

You cannot beg or make one love. Not a true love. One has to come to love on one's own accord: Most start with their parents first; then usually another common gene carrier like a sibling or grandparent. Later, perhaps, a significant other or dear friends. But ultimately, and hopefully, oneself, and all life.

None should go unloved. Love is the answer for universal peace. Universal peace is the answer to suffering. Combatting suffering and deriving order from chaos is one of the meanings of life. This is God's will. It is possible to love everyone and everything. This must come from a visceral sentiment founded in each individual's mind.

Loving someone you share genes with is biologically easier, but no less meaningful. Loving someone outside of your gene pool is more complex, but very possible and very liberating to express unconditionally. Free love is freedom of mind and freedom of consciousness. Love is the answer.

Hope

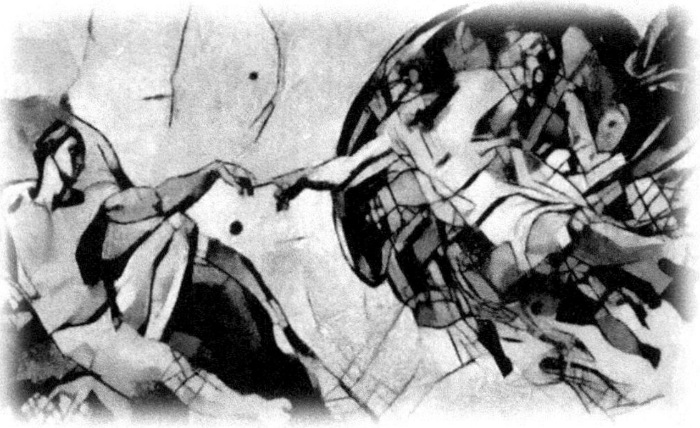

Hope is a powerful type of mindful conscious execution which many people have engaged with but do not truly understand. Hope is akin to prayer; the energy it invokes is in line with karma. In language hope is a qualifier of intent via admittance that a given outcome is out of your control. Intent is mindful directed energy.

Hope falls between want and demand. The beauty of hope is that it does not want or demand; it only expresses to the universe the state of one's intent. You must not beg anything back of the universe; you must simply allow yourself to have the thoughtful intent of *being* hopeful. Hope is a state of being.

Humans only hope for outcomes that are outside of their control; otherwise they would control the outcome. Hoping for something you can control is to express laziness and ego. Hoping for something you cannot control is natural. Accepting you cannot control the outcome is true hope. Be honest with yourself.

True hope may only be experienced and expressed. The complement of chance and certainty is action, motion, and decision. True hope is acceptance of the uncertainty of outcomes which emerge across the backdrop of infinite certainties. This provides deep meaning for life and the observer.

Through hope, your consciousness maintains a balance of reciprocity without unfair despair. If you want or demand an outcome you are trying to persuade karma and this can lead to shifts in balance and peace of mind. One, who hopes, not demands, isn't burdened by any outcome: Peace through hope and ego death.

If you were to beg something back of the universe then this is not hope, this is a want or a demand. The universe has no responsibility to fulfil any of these, but by aligning your intent through a state of being of true hope, accepting the uncertainty of outcomes, the universe's response will be that of karma.

Karma may not manifest the reality you hoped for, but if it was a true hope your mind and brain will at least be happy at peace. If it was a want or a demand then your mind and brain will not be at peace. Your mind and brain inherit the imbalances of your ego's unmet desires and failed acceptance of outcomes.

The imbalance inherited by your mind and brain will lead you to experience destabilizing emotions, sensations, and behaviours. You force this on yourself when you demand an outcome which is incongruent with reality. The mismatch between your mind's intent and reality leads to cycles of chaos and order.

There are many matters of things that you must accept you have no control over. Accepting this will bring you happiness through peace. Hope remains when all else seems lost. Hope is also there in times of prosperity. This is the nature of hope. True hope aligns one's reality; it cannot be wielded, only expressed.

Peace of mind can be achieved by learning to hope and accept reality instead of yearning desires. Hope invokes balance and fair order. Fate is not mere happenstance nor is it entirely determined. The conscious universe can always intervene. Thus, don't demand, be honest with yourself and hope for the best.

Death

Biological life forms eventually die, but their matter and energy remains in the universe. In life, energy is not lost; it transforms. Energy cannot be created or destroyed but it may change states: First law of thermodynamics. As such, death is merely a transformation of energy from one state to another energy state.

Biological death, in the animal kingdom, is a word we use to describe the decay and failure of our bodies' cells. Our bodies are just densely formed carbon based systems. Some animals' bodies won't decay for hundreds of years, whereas others decay in a day. Presently, the oldest recorded human body lasted 122 years.

If human bodies could live forever, or even for 1,000 years, our perspective of death would be drastically different. It is the juxtaposition of finite living, against the apparent infinite backdrop of death which creates the anxiolytic notion of impermanence. Eliminate bodily decay, and eliminate this connotation of death.

Silicon and AI based life may overcome decay and therefore defeat biological death. Through ease of repair, replacement, and sustainability silicon may not only be more suitable for longevity but also adaptability. Current biological life forms have many limitations which restrict us to our present environments.

Biological life is fragile yet abundant. Each species evolved to thrive in its own ecosystem. A fish cannot thrive on land and a human cannot thrive underwater. We have evolved to take advantage of our present environments. Intelligently designed AI bodies could potentially overcome any environment.

Theoretically, humans should also be able to overcome any environment, including our own biological bodies. This is done in practice every day. By the use of human machine cultivation we have explored the skies, the seas, the land, and space. Further, we use medicine to save bodies which become wounded or ill.

We must address death as a disease which is combatable. This includes all diseases which ail biological life, including aging itself. New anti-aging research is offering drastically improved longevity of biological bodies. Our ability to manipulate genetics and use nanotechnology will allow us to stop or even reverse aging.

Neurotechnology may even enable a human upload into a new biological or AI system. In other words, we may eventually transcend biology entirely. You are now alive in a biological body, but you were not always. Before you were born, you were not dead. Life and consciousness is more than biology.

The lived experience of bodily disease and decay is called suffering. Death is one path to resolve suffering: Science, technology, medicine, meditation, peace, and love are another. Human bodies will decay over time, but our imprint on universal consciousness will remain in the fabric of nature.

Biological death means the end of your body. Your energy will continue to exist and may transform into something more harmonizing than biological life itself. Therefore, enjoy your energy life force in its present form and present body before it is transformed into a new experience. Death is not the end.

That’s life.

I would love to hear from you!

www.MattBuchananStudios.com
FB-YouTube-Apple-Spotify-SoundCloud-
GooglePlay: Matt Buchanan Studios
FB and YouTube: Matt Buchanan Comedy
Twitter: BuchananNeuro #AEFL
Twitter: Matt_Comedy
IG: Matt_Comedy
IG and TikTok: PoliteDiscoveries
IG: CanadaNeuroInc
www.CanadaNeuro.com
ResearchGate.net/profile/Derrick_Buchanan
LinkedIn.com/in/DerrickMatthewBuchanan

Please subscribe to one of my channels!

Including my: scientific publications and presentations, science/comedy/arts related podcasts, social media broadcasting channels, and business enterprises.

Author Bio

Derrick Matthew Buchanan is a Canadian Neuroscientist, Business Owner, and Comedian from Ottawa, Ontario. He has a PhD in Neuroscience from Carleton University, a Bachelor of Cognitive Science with a Specialization in Cognition and Psychology, and has also obtained certificates from McGill Montreal Neurological Institute, and Harvard Medical School. Dr. Buchanan also completed a PhD fellowship at the University of Geneva Switzerland and has been the recipient of many prestigious academic and research awards. His research applications are primarily focused on the use of neuroimaging, brain stimulation, brain computer interfaces, and machine learning to improve the classification and treatment of mental health and neurological diseases. As a clinical neuroscientist he has authored dozens of scientific publications and academic essays.

Dr. Buchanan is the Founder and CEO of CanadaNeuro Inc, a clinical research institute for mental health and neurological diseases. He is also an avid consumer of physics and enlightenment literature including ancient and modern teachings from science and mathematics to philosophy and religion. Dr. Buchanan is also an avid consumer of all varieties of live and recorded art, music, theatre, magic, and comedy. Ultimately he is a person who lives to see ideas come to life. He says that taking ideas from your brain and consciousness and manifesting them into reality is one of the fundamental reasons for living. He reminds us that everything we create is a product of consciousness.

Although he is an active neuroscientist, he founded www.MattBuchananStudios.com as an opportunity to create a business out of his favourite hobbies. This includes writing books, articles, and scripts; but also writing and performing stand-up comedy, producing his podcast and YouTube channel Matt Buchanan Studios, and directing and producing films.

Other Written Works by DM Buchanan

Scientific Articles

Buchanan, D.M.; Bogdanowicz, T.; Khanna, N.; Lockman-Dufour, G.; Robaey, P.; D'Angiulli, A. (2021) Systematic review on the safety and tolerability of transcranial direct current stimulation in children and adolescents. *Brain Sciences. Vol. 11, 212.* DOI:10.3390/brainsci11020212

Buchanan, D.M.; Robaey, P.; D'Angiulli, A. (2020) What Do We Know about Transcranial Direct Current Stimulation for Major Depression? *Brain Sciences. Vol. 10, 480.* DOI:10.3390/brainsci10080480.

Buchanan, D.M.; D'Angiulli, A.; Samson, A.; Maisonneuve, A.R.; Robaey, P. (2020) Acceptability of transcranial direct current stimulation in children and adolescents with ADHD: The point of view of parents. *Journal of Health Psychology* *DOI*:10.1177/1359105320937059.

Buchanan, D.M; D'Angiulli, A.; Samson, A.; Robaey, P. (2019) Transcranial Direct Current Stimulation in Pediatric Neuropsychiatry: Safety Tolerability and Acceptability. *The Journal of Psychiatry and Neuroscience*.

Buchanan, D.M.; Croarkin, P.; Cohen Kadosh, R.; Kirton, A.; Masuda, F.; Noda, Y.; Gilbert, D.; Donaldson, P. (2019) Advances in Pediatric and Neurodevelopmental Brain Stimulation. *International Brain Stimulation Conference Symposium*.
DOI: 10.13140/RG.2.2.12081.02408

Buchanan, D.M.; D'Angiulli, A.; Samson, A.; Amare, S.; Gaumond, G.; Robaey, P. (2019) Making transcranial direct current stimulation treatment in atypical child and adolescent neurodevelopment a reality: Translating safety tolerability and acceptability evidence from the laboratory into the doctor's office, the classroom, and home. *Brain Stimulation*. 12(2):474. DOI: 10.1016/j.brs.2018.12.545

Buchanan, D.M.; Grant, J.; D'Angiulli A. (2019) Commercial Wireless vs. Standard Stationary EEG systems for Personalized Emotional Brain-

Computer Interfaces: A Preliminary Reliability Check. *Neuroscience Research Notes.* 2(1): 7-15. DOI: 10.31117/neuroscirn.v2i1.21

Piexto, C.; Hyland, L.; **Buchanan, D.M.;** Langille, E.; Nahas, R. (2018) The Polytrauma Clinical Triad in Patients with Chronic Pain after Motor Vehicle Collision. *Journal of Pain Research.* 11:1927-1936. DOI: 10.2147/JPR.S165077

Buchanan, D.M. (2013) Sound Changes of the /h/-phoneme from Old English to Modern English. *Canadian Undergraduate Journal of Cognitive Science*; 47-54.

Online Essays

Buchanan, D.M. (December 2013) The Evolution of Rational and Irrational Thinking. *Cognitive Evolution.* DOI: 10.13140/RG.2.2.31745.51046/1

Buchanan, D.M. (December 2013) What makes language so special? – A short summary of the stimulating debate. *Language and Cognitive Science.* DOI: 10.13140/RG.2.2.33423.23204

Buchanan, D.M. (November 2013) The Piraha Tribe: Chomsky vs Nevins on the Implications of Cultural Constraints on Natural Language Development. *Language and Cognitive Science.* DOI: 10.13140/RG.2.2.16803.30249

Buchanan, D.M. (April 2013) Confidence Effects on Rule-Based and Information-Integration Category-Leaning. *Cognitive Psychology.*

Buchanan, D.M. (February 2013) The Effect of Presentation Order on Function-Learning Tasks. *Cognitive Psychology.*

Buchanan, D.M. (December 2012) The Effect of Source Credibility on Syllogistic Reasoning. *Cognitive Psychology*.

Buchanan, D.M. (November 2012) The Effect of Reading Ability on The Attentional Blink. *Cognitive Psychology.*

Buchanan, D.M. (December 2011) The Predication of Existence. *Philosophy of Language.* DOI: 10.13140/RG.2.2.33449.44645

Buchanan, D.M. (October 2011) Frege's Puzzle Substitution Salva Veritate. *Philosophy of Language*. DOI: 10.13140/RG.2.2.30067.78889

Buchanan, D.M. (April 2011) Language and Identity: How Language Necessitates and Creates Identities. *Applied Linguistics and Discourse*. DOI: 10.13140/RG.2.2.20027.67363

Buchanan, D.M. (April 2011) The Chinese Room Thought Experiment and the Robot Reply. *Philosophy of Mind*. DOI: 10.13140/RG.2.2.35100.95364/1

01000001 00100000 01110011 01101001
01101110 01100111 01110101 01101100
01100001 01110010 01101001 01110100
01111001 00100000 01101001 01110011
00100000 01101110 01100101 01100001
01110010 00100000 00101101 00100000
01000100 01001101 01000010

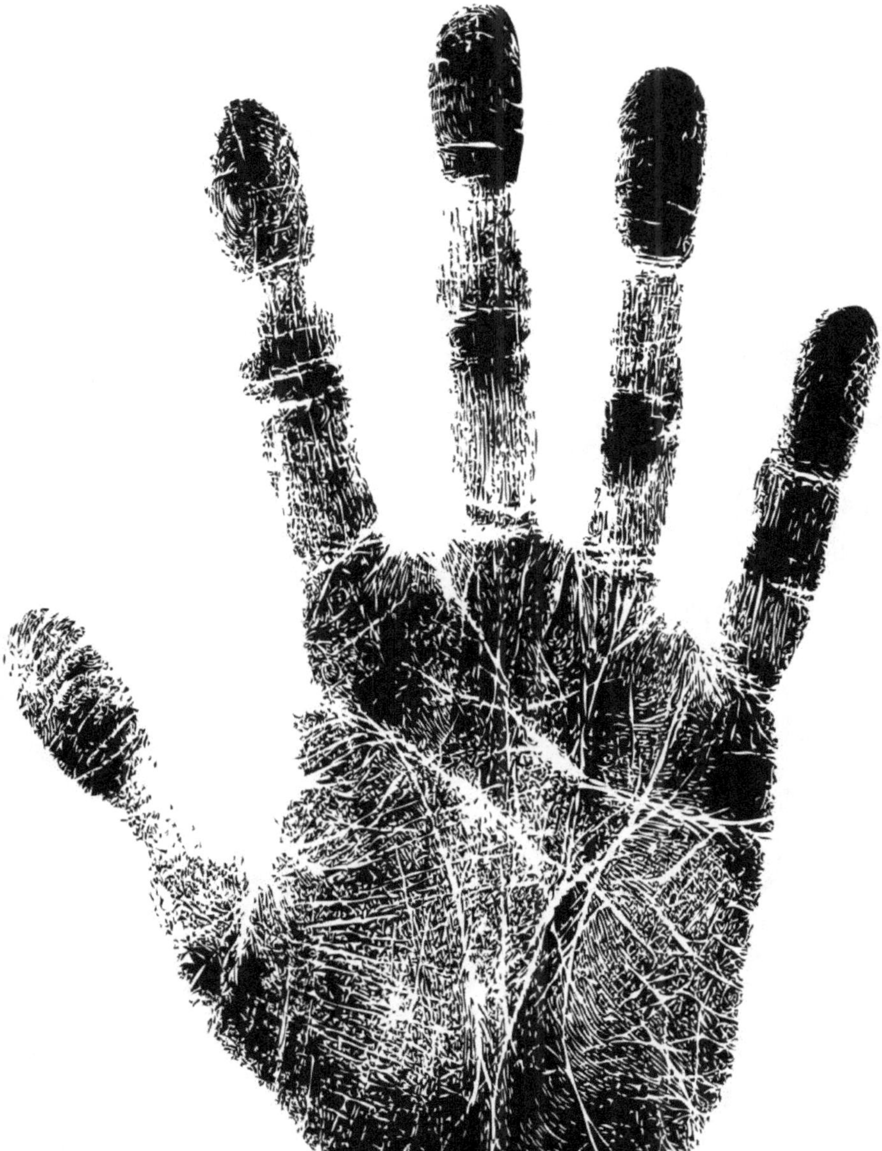

www.ingramcontent.com/pod-product-compliance
Lightning Source LLC
LaVergne TN
LVHW010722061126
851857LV00014B/1301
* 9 7 8 1 7 7 7 1 5 8 4 0 8 *